AMERICAN RUSTIC

Classic Barns

American Rustic

Classic Barns

April Halberstadt

MetroBooks

MetroBooks

An Imprint of the Michael Friedman Publishing Group, Inc.

Library of Congress Cataloging-in-Publication Data

Halberstadt, April, 1943–
  Classic barns / April Halberstadt
    p.cm. – (American rustic)
  Includes bibliographical references and index.
  ISBN 1-58663-176-4 (alk. paper)
    1. Barns—United States. 2. Vernacular architecture—United States. I. Title. II. Series.

NA8230 .H285 2001
728'.922'0973—dc21

2001030819

Editor: Daniel Heend
Art Director: Kevin Ullrich
Designer: Lissi Sigillo
Photography Editor: Kathleen Wolfe
Production Manager: Rosy Ngo

Color separations by Fine Arts Repro House Co. Ltd.
Printed in China by Leefung-Asco Printers Ltd.

1 3 5 7 9 10 8 6 4 2

For bulk purchases and special sales, please contact:
Friedman/Fairfax Publishers
Attention: Sales Department
230 Fifth Ave
New York, NY 10001
212/685-6610 FAX 212/685-3916

Visit our website:
www.metrobooks.com

ACKNOWLEDGMENTS

Charlene Duval; Hans Halberstadt; John Olson of Barn Again!;
Elizabeth Stephens; Mike Shepard; Dennis J. Pogue; Eric Gilbertson.

To Chris and Stephanie Hope

CONTENTS

INTRODUCTION

The incense of seasoned wood and perfume of dry hay mingled to create that distinctive fragrance which only an ancient barn possesses.

— Eric Sloane, *An Age of Barns, 1967*

An American barn is one of a handful of rare symbols that immediately evoke our national heritage. Red, white, and blue with the stars and bars of our flag; a magnificent bald eagle with soaring wings outstretched—these are both icons that symbolize America. Following closely is the image of a barn, another enduring symbol of the American pioneer spirit, and one that recalls the hard work and cultural values that combined to build our nation. The barn is also a reminder that we are a nation of immigrants, and of entrepreneurs.

As America developed, our forefathers began building structures that featured a diverse array of decorative styles. Houses were built to copy Roman temples or Gothic cathedrals. Barns, however, stayed honest and unpretentious. If ever there was an example of "form follows function," it is our American barn. Made of local materials by local craftsmen, barns were built to fit the landscape and serve the needs of the farmer.

In early America, every man was a farmer. Although many pioneers were capable carpenters or blacksmiths, these occupations were usually carried out in addition to farm work, and, despite their many skills, American colonists were not entirely self-sufficient. Early America was overwhelmingly rural, and the advent of the barn—most built larger than the family homes they shared land with—was what allowed many farmers to gain a good degree of working independence. In this way, barns might be considered America's first industrial buildings.

No two barns are exactly alike, which is another reason for their appeal. Americans take special pride in being singular individuals, and barns reflect that uniqueness. Barn design changed as America grew, due in great measure to evolving farm practices, but also due to technological innovations. Barns also reflect local crops, customs, culture, and weather. In this era of "global village" commerce and communication, it is our vernacular buildings, our barns and other local structures, that help us retain our regional identity.

"Vernacular architecture" is the term architectural historians and urban planners use to describe simple, locally designed structures that reflect regional traditions. It refers to everyday, down-to-earth, do-it-yourself buildings. Farm buildings, barns, sheds, corn cribs, silos, and outhouses all fall into this classification. Once overshadowed by high-style, architect-designed estates and overlooked by scholars as inconsequential, vernacular buildings are now being rediscovered, restored, and appreciated.

Hundreds of books are written each year profiling historic houses, yet few are written about barns. Over the years, much has been written about farm practices and methods of production, but few books discuss the design and construction of the farmer's most important tool, his barn. A significant body of information about old barns is available from a somewhat unlikely source: old farming magazines. Barn plans and construction methods were frequent features in farm periodicals dating back to the Civil War era, and they provide us with some of our best clues when we try to analyze the age and construction of an old barn.

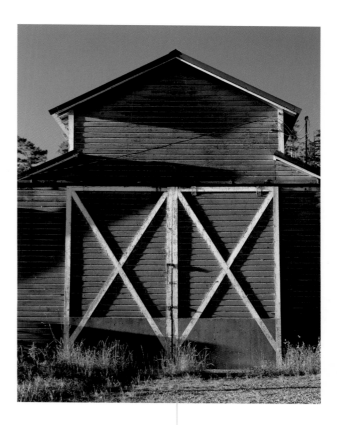

ABOVE: **The tall, generous doors indicate that this barn was built to accommodate a very tall vehicle.**

While barn design reflects the local weather and available building materials, it also reveals the nature of the crop: Is it wheat or corn? Does the farmer keep cows and horses? Was this a dairy? All these important considerations have shaped barn design at one time or another. And these are all clues to figuring out the age and the nature of an historic barn.

LEFT: **Because of the diversity of traditional barn architectural styles—and the advances in restoration technology—it's difficult to tell if this barn is one month old or one century old. It is a stately and dignified structure, whatever its age.**

BELOW: *Built by local craftsmen using materials from the area, barns are classified as vernacular architecture. Frequently overlooked, it is the local architecture that gives a region its unique character and personality.*

As the world becomes more and more modernized, a growing segment of the population has developed an interest in barns and other vernacular architecture. Since fewer families are living on farms and many old barns are reaching the end of their useful lives, these beloved American icons are disappearing rapidly. Enthusiasts are beginning to realize the importance of these old buildings, and thus the best new books about barns are usually small, scholarly studies done through local universities. These studies can provide the most thorough research about rural architecture in specific regions.

The majority of Americans now live in cities, and few new barns are currently being built. Only the Old Order Mennonites and a few related groups who still practice traditional farming continue to build

barns. But many of the rest of us love barns and go to absurd lengths to keep their legacy alive. The word "barn" now appears in the names of hundreds of urban businesses. Dress shops, used car lots, and antiques stores all use the word in their names, marketing the idea that barns are warm, friendly, comfortable places.

At the turn of the nineteenth century, about half the American population lived on farms. Today, a hundred years later, less than two percent of Americans list farming as their occupation. Farming has become so efficient that a very small percentage of the population can feed the entire nation—and still produce crops for export. Many modern farm operations no longer need a barn. Instead, silage is kept in the field, stored in huge black plastic casings that line the field like sausages, to be scooped out and fed to the cows with a front-loading tractor. The need for speed and efficiency has all but eliminated the use of barns and silos.

Still, barn appreciation has come into its own in the last two decades, as more and more people have begun to recognize the value and importance of vernacular buildings. Just a few years ago, barns were

being demolished to salvage the weathered wood for decorative purposes. Recognizing the value of the old timbers—some of them hand-hewn—architects and craftsmen frequently dismantled barns in order to preserve the materials. Many barns were also modified into country houses in an attempt to save them, thereby losing their functional capability but maintaining some of the history. Today, tremendous efforts are being made to preserve the dwindling number of barns that remain. All across America, programs are now in place to assist barn owners in rehabilitating their structures by promoting their usefulness as . . . barns.

During your next drive in the country, enjoy the barns that remain. Many of them tell a little story that is obvious to the knowledgeable observer. A barn can recall the basis of the local economy, reveal the time period that a community developed, and indicate how wealthy and productive the area became over the years. A barn does not have to be large or elaborate to be historically significant. Listen to the stories that barns tell us. They are telling us about ourselves.

OPPOSITE: In the past, barn sites were carefully chosen before building was begun. Adequate water for livestock was an important consideration, but so was good drainage.

ABOVE: The presence of a barn owl is considered to be a good sign, the mark of a prosperous farm. Barn owls eat mice and other rodents, providing essential vector control without dangerous poisons.

LEFT: Barns were important to early American businesses, advertising everything from animal feed to liniment to chewing tobacco. Farmers were sometimes given a small fee for this advertising space, but others were happy just to have their barns painted.

COLONISTS AND YANKEES

Chapter One

Agriculture has ever been amongst the most favorite amusements
of my life.

— George Washington to Arthur Young, August 6, 1786

One of George Washington's favorite agricultural "amusements" was his barn, a sixteen-sided structure built from his own design. We think of Washington as our first president, commander of the Revolutionary Army, and the father of our country. But visitors to his home in Mount Vernon are surprised to discover that Washington was also a leader in farming—a plantation owner who experimented with many of the new farming theories of his time and developed a few of his own.

By the time Washington was developing his farms, agricultural practices in America were quite sophisticated. The farmers of Washington's time knew which types of crops were suitable for American climates and soils and

LEFT: *Built above ground level, the raised floor of this barn allowed horses and wagons to be driven into the upper level. The manure and used bedding straw were shoveled into a wagon on the lower side.*

ABOVE: *Windows all along the lower level indicate that this structure was probably initially built as a dairy barn.*

BELOW: *The earliest barns tended to be simple structures, usually built by the enterprising farmer with little outside help.*

which would fail. They knew the benefits of manure and experimented with different types. By George Washington's day, American farmers had already accumulated 150 years of farming experience, and had built barns and other agricultural buildings with designs that reflected that experience and knowledge.

The Colonial Farmers

America began as a nation of colonists, a brave group of settlers who had been sent across the Atlantic Ocean by wealthy investors looking to exploit an uncharted land. While many records exist concerning the arrival of the first European settlers and the appearance of the shelters they built, little specific information remains about their barns. Thus, finding and verifying the date of the earliest barn in America is an impossible task. We know that the Eastern seaboard features many styles of vernacular architecture, and that these reflect the customs and heritage of the early settlers, as well as the local weather conditions and the types of crops produced. So we can assume that the early barns were small, low buildings with thatched roofs.

Perhaps the earliest documentation about farming practices and barns comes from the letters that William Bradford, the first governor and historian of the Plymouth colony, wrote around 1629. Bradford's journal, *History of Plymouth Plantation, 1620-1647*, is our first look at early American farming practices.

We know that there were no farm animals transported on the *Mayflower*, but it quickly became apparent to the settlers that livestock would be needed. Farm animals were soon imported for the colony and, in 1627, Bradford wrote about the division of agricultural assets in the community, mentioning cattle, goats, and pigs. So we know there were substantial numbers of animals in the colonies by that time.

OPPOSITE: *While this West Virginia barn's saltbox shape was most common in New England, examples also occurred in the South, where the roof was sometimes referred to as a "catslide."*

Oxen were the most common draft animal, and dairy cows were also kept by the early colonists. Draft horses did not appear in great numbers until the later years of the eighteenth century; oxen were less expensive and easier to keep. We know that some sort of barn or shelter was used to protect the animals from predators at night and in bad weather.

FOLLOWING PAGES: *We might think of the early barn as a food-processing factory, the structure where farm products were initially gathered for future distribution. Large farms usually maintained at least two barns—one for crop storage and threshing, and one for livestock.*

Early settlers hoped to grow the same crops they were familiar with from the old country. However, it quickly became apparent that climate conditions in America were quite different from those in Europe, and that the colonists would need to adapt their traditional farm practices to suit the new land. There was also a tremendous interest in growing some sort of crop for export and, by 1620, tobacco-growing had become widespread in the area of Jamestown, Virginia.

RIGHT: **Early seventeenth-century tobacco barns in the Chesapeake Bay area were long, low sheds. Later examples such as this one near Weston, Missouri, look more like conventional barns.**

BELOW: **Tobacco sheds tend to be used for only one crop and do not lend themselves well to other farm uses. This is due in part to the specialized system of interior framing used for hanging and drying the tobacco leaves.**

Tobacco Barns of the First Colonies

Historians frequently note that the types of crops grown in early America had a profound influence on local social relations and cultural values. For example, tobacco was grown in the Chesapeake Bay area, mixed grains were harvested in the middle colonies, and rice and later cotton were grown in the South. The farms and the barns of these regions gradually began to reflect the economic base of each area. So, in addition to the little barns needed to protect a variety of animals and stored feed during the harsh winter, the Chesapeake Bay area features one of the first specialized crop barns built in America: The tobacco shed.

Tobacco sheds are a very distinctive type of barn, easily identified by their long, low shape and the placement of ventilation openings. These barns were created solely for the drying of leaves that were used for cigar wrappers. Illustrator-historian Eric Sloane became intrigued by these peculiar structures early in his career and reported that he painted pictures of them exclusively for many years. His initial studies of old tobacco barns led him to a lifelong interest in and study of other old farm buildings.

ABOVE: *Tobacco was America's first cash crop, and thus tobacco sheds were once a common sight in eastern states. A tobacco barn is easily identified by the ventilation spaces between the boards of the siding, as well as by the interior framing on which the tobacco leaves are hung out to dry.*

Tobacco barns are unique because they are the earliest examples of specialized industrial buildings in America. Built with a system of internal cross members that allowed racks of tobacco leaves to be hung for air-drying, the shape, interior framing, and ventilation systems of the tobacco barn prevented it from being easily reused for other agricultural purposes.

Tobacco culture in America began as early as 1612; thus, the appearance of specialized barns in America dates from the earliest colonial times. While the earliest tobacco culture took shape around the Chesapeake Bay, tobacco barns can still be found along the eastern seaboard all the way up into Connecticut.

MAIL POUCH TOBACCO

AMERICA'S MOST FAMOUS BARN PAINTER, Harley Warrick, died late in the millennium year, 2000, and his legacy is now fading. He was the last member of an elite crew of barn painters. For fifty-five years, Harley painted signs promoting Mail Pouch Tobacco, a product advertised on hundreds of American barns during the twentieth century. "The first thousand were a little rough, and after that, you get the hang of it," he noted.

One of the more colorful local customs throughout agricultural America was to have your barn painted by an advertiser, and for years the most famous and the most pervasive advertising on barns was developed by the Bloch Brothers Tobacco Company of Wheeling, West Virginia. Their slogan "Treat Yourself to the Best, Chew Mail Pouch Tobacco" was found on hundreds of barns from Florida to Oregon and all points in between. Most Mail Pouch ads are black with blue and white lettering.

Mail Pouch chewing tobacco was developed in the late nineteenth century; the Bloch brothers began advertising on barns in the 1890s. They originally contracted with local painters, but by the 1930s the Bloch brothers had begun to work with regional contractors. Then the program was consolidated into one company that picked the barns and managed the painting teams. The painting teams themselves usually consisted of two men who were sent all across the country adding advertisements to barns. It took a full day to paint the side of a 40-foot (12.2m) barn, but since the design itself was standard, the most difficult task was sizing the letters to fit the barn in question.

Sites selected for painting had to meet some basic but critical guidelines. The sign needed to be visible from the highway but had to be situated so drivers could read the sign without taking their eyes off the road. There could be no windows that might interfere with the lettering. The farmer was paid a small annual fee for the advertising, although, in later years, the presence of a Mail Pouch sign became such a status symbol that farmers even offered to pay the company for the privilege.

In 1966, when the Highway Beautification Act banned commercial advertising 600 feet (182.9m) from interstate highways, the Mail Pouch advertising program went into decline and the company retained only one team of painters to maintain their signs. People lobbied to save the picturesque signs, and Congress declared an exception to the Highway Beautification Act in 1974.

Tobacco billboards faced another challenge in April 1999, when Congress forced the cigarette industry to take down outdoor advertisements for tobacco products. The agreement was part of the $206 billion federal tobacco settlement.

Through years of controversy, through the decline of tobacco advertising, the elimination of lead-based paints, and the debate over roadside advertising, the Mail Pouch signs have prevailed. They have been maintained by barn owners who love them and are willing to preserve a unique American rural tradition that is over a century old. Harley retired in 1993 but continued touching up his handiwork until a month before his death. Someone else will have to carry the tradition into the twenty-first century.

Painted signs such as this one are now targeted for razing due to federal prohibitions on tobacco advertising. Preservationists are working to save these unique signs from destruction.

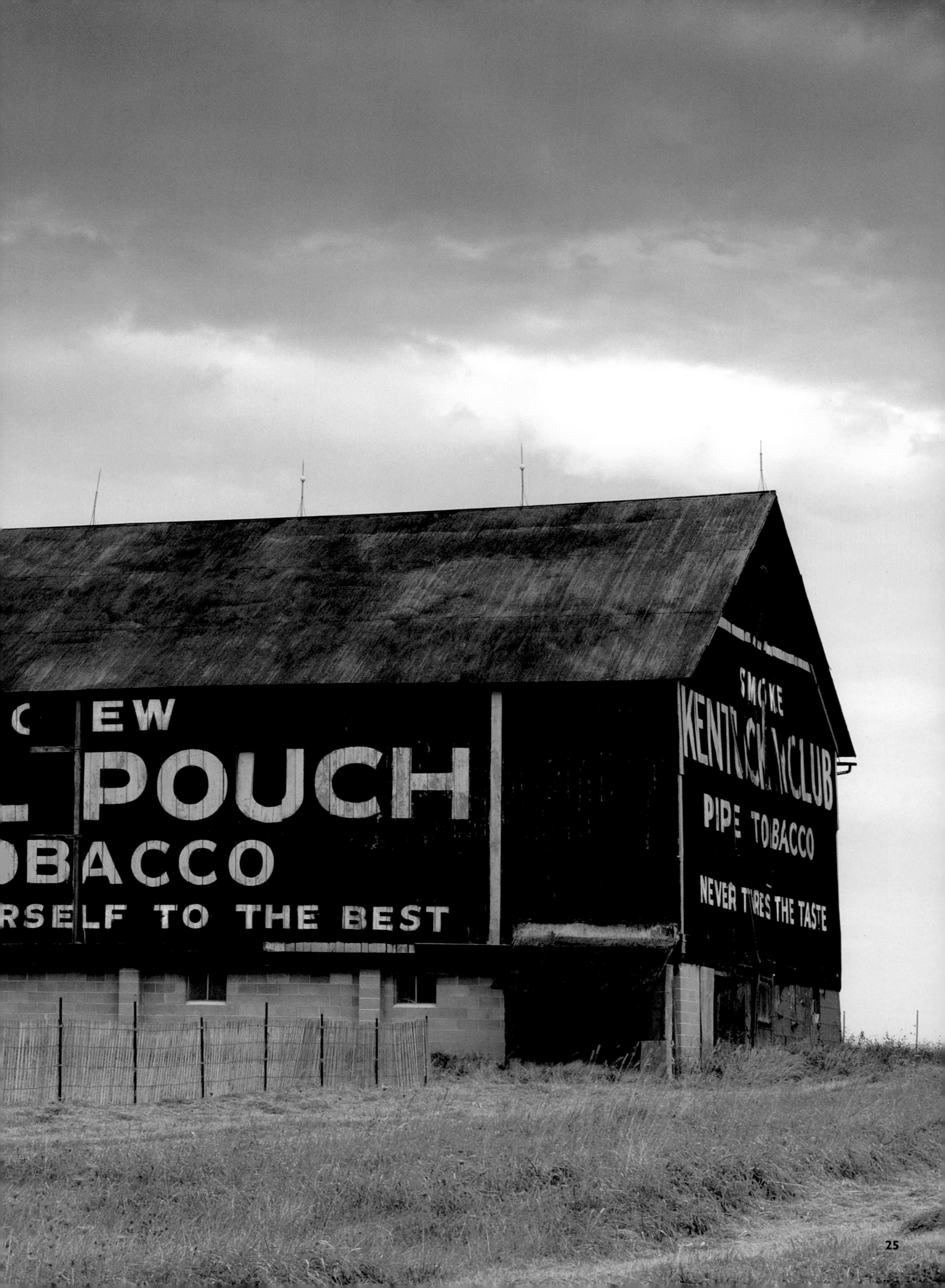
POUCH
OBACCO
RSELF TO THE BEST
SMOKE
PIPE TOBACCO

RIGHT: *The eastern United States has many fine examples of stone barns. Especially prized, they frequently make wonderful living quarters if carefully adapted.*

BELOW: *An affection for barns seems to be universal. The tradition of building a red barn with white trim and a crowing rooster weather vane has great appeal in rural areas.*

The English Barn

The English barn, sometimes called the three-bay barn or the Yankee barn, is undoubtedly one of the most satisfying architectural shapes in existence. Indeed, when young children draw pictures of a barn, this is the structure that they tend to create: a large shed with a generous door precisely in the middle. In actuality, the middle door of the English barn is large enough to accommodate a wagon loaded with crops, and hay can be thrown into the lofts on either side.

It can be argued that the appearance of the English barn in America marks the beginning of our independence from European farm design, since the small barns with the thatched roofs were now gradually being replaced by this new design. One of the most complete descriptions of the English barn comes from a travel diary written by Peter Kalm in 1748. Kalm traveled through Pennsylvania and New Jersey, all the while noting the differences between the farms and farm practices in the New World and the Old. He observed:

ABOVE: **Red is the traditional color used for barns, and is derived from mixing inexpensive iron oxide with linseed oil and turpentine. It was a coating that covered walls well and proved to be quite durable.**

> *The barn had a peculiar kind of construction hereabouts, which I will have a concise description of. The whole building was very great, so as to almost equal a church; the roof was pretty high, covered with wooden shingles, inclining on both sides, but not steep: the walls which support it were not much higher than a full grown man; but, on the other hand, the breadth of the building was more considerable: in the middle was the threshing floor, and above it, in the loft or garret, they put the corn which was not yet threshed, the straw, and any thing else, according to the season; on one side were stables for the horses, and on the other for the cows. And the small cattle had likewise their particular stables or styes; on both ends of the building were great gates, so that one could come in with a cart and horses through one of them, and go out at the other. There was therefore under one roof the threshing floor, the barn, the stables, the hay loft, the coach house, etc.*

The English barn is a relatively common structure and appears widely throughout the eastern half of America. It is usually of modest proportions, appears to have been simple to build, and was easily adapted to local farming practices.

George Washington's Sixteen-Sided Barn

The father of our country, George Washington, was an extraordinary farmer. With four working farms on his 8,000-acre (3240ha) estate, his plantations occupied most of his attention. Washington was also a thoughtful and pragmatic agriculturist, always looking for ways to improve his crops and make his plantations more productive. Fortunately he kept a diary of his farm endeavors, so we have a good record of both his theories and his practices.

Washington and his wife, Martha, started with the home site on the Potomac River that had belonged to George's grandfather. The Widow Custis, as Martha had been known, brought substantial acreage with her when she married Washington. With such a large farming operation, managing, harvesting, and storing the sizable crops presented problems of tremendous magnitude.

Like many other planters in Virginia,

ABOVE: **Just another step in the crop cycle: The grain is cut and laid to dry in the field.**

Washington read the latest books and magazines from England. Interested in progressive farming techniques, he corresponded with men like Arthur Young, a progressive English farmer noted for his scientific farming methods. Washington imported new seeds, experimented with various types of manure, and analyzed soil conditions in order to match each crop to its optimal location on his farms.

Always interested in improvement, Washington searched for various ways to thresh using less labor. He used slaves to supply the labor on his plantation, but this practice troubled him greatly. His solution was inspired in part by a neighbor, John Beale Bordley, who had written a treatise on treading out wheat using animals. This method of threshing calls for the animals to walk on the harvested grain, allowing pressure from their hooves to separate the kernels from the stalks.

After consulting with Arthur Young, Washington built a new barn for threshing on his Dogue Run farm. Since there was a convenient source of brick clay nearby, the new barn was built of brick. It was a substantial structure with enough room on the threshing floor for thirty men to flail the crop.

Wishing to use horses for threshing, Washington designed a new barn, this time utilizing a roughly circular design. Once again he used bricks to build his new threshing barn, but this structure was sixteen-sided and about 52 feet (15.8m) in diameter. It had two floors: a threshing floor with gaps between the boards on the upper story and a granary below. The horses would tread in a circle, threshing the grain with their hooves, and the small kernels of grain would drop between the boards and land on a solid wooden floor on the lower level. Straw and animal droppings were quickly removed by workers on the upper floor.

This process offered some tremendous advantages, since threshing could now be an all-weather operation, allowing grain to be processed at any time during the year. The granary on the lower floor was well ventilated yet secure, as Washington was often away and concerned with theft of his crops.

Completed in 1794, the original Washington round barn remained on his farm for nearly a century. The barn apparently did not meet expectations as an efficient way to thresh grain, though, as keeping the grain clean and free of animal residue posed a significant problem. In addition, other technologies soon arrived, improving grain harvesting enormously. The portable thresher, a machine developed and perfected by J. I. Case, was in the fields in the early nineteenth century, and immediately eliminated the need for hand threshing.

Washington's round barn deteriorated and then disappeared in the 1870s, but in 1996 a reproduction of the barn was built on the original site. Every attempt has been made to replicate the original barn, even to the manufacturing of hand-forged nails to attach the siding and floors. The centerpiece of an ongoing exhibit, the new barn is the result of five years of effort by the Mount Vernon staff, consultants, interns, and volunteers. The bricks, shingles, and other construction materials were carefully chosen to match the original as closely as possible.

BELOW: *George Washington's round barn is unique in several ways; the construction materials alone make it interesting to historians. It was originally designed as a threshing barn and is a radical departure from the three-bay threshing barns with large doors in the center that were commonly built at the time.*

George Washington was interested in two other significant agricultural innovations, but only one would have lasting influence on barn design. Washington tried to introduce the hedgerow to America to replace fences, but this undertaking was met with very limited success. His other introduction was much more important, though he is seldom credited for playing a significant role in its inception. Washington introduced the mule to America, importing two jacks and two jennies to his farms. They were a gift from the king of Spain, and Washington was so impressed with them that he hoped to have his official carriage pulled by a matched team of mules.

The introduction of the mule would have an eventual impact on barn construction, since the stalls in a mule barn can be much smaller than those designed for horses. The introduction of the jackass and its eventual breeding for mules would have an enormous impact on southern agriculture too. Many small southern farmers depended on the mule for their livelihood well past the middle of the twentieth century.

ABOVE: *In the cold northeastern states, connected barns are a common sight. These structures place the house, barn, silo, and other farm structures in a u-shape so the farmer does not have to go out in cold, bitter weather to feed his livestock.*

The Connected Barn or Maison Bloc

One of the earliest and most practical farmstead configurations is the connected barn, wherein the barn and the farmhouse share a common wall. This type of barn, or rather farmyard arrangement, is very common in the New England area and in parts of Canada. It is a type of vernacular building that appears very early in the development of North American architecture; some examples date to the seventeenth century. Where winters are severe, the farmer can feed and tend to his livestock without venturing out into the snow.

The connected barn is an enclave of buildings, each designed for a different purpose. Sometimes the buildings are arranged in a square around a central courtyard. Other times the separate areas for livestock, hay storage, silage, the grain bin, and pens for the various farm activities are located in a row, one after the other under a single roof, with common walls between them.

In the upper reaches of the continent, where immigrants from Brittany settled in Canada, the block house, or *maison bloc*, is a common sight. Although it looks like a large and very comfortable house, with everything under one roof, the family is actually housed in one end of the building, and the livestock and their fodder occupy the rest of the structure. The roofline gives no clue about which end is which. Both the connected barn and its cousin, the maison bloc, appear very early in the agricultural areas of the New World. While there are similar structures in Europe, most of the North American buildings appear to have been modified to suit local conditions.

LEFT: **Sometimes several stories tall, each level of the barn has a unique function. Grain is stored on the top level, and is threshed and fed to the livestock on the middle level. The resulting manure is thrown into waiting wagons on the bottom floor.**

BELOW: **The maison bloc, or block house, is a common type of barn in northern Canada. One roof covers both the barn and the farmhouse, sometimes making it difficult to distinguish between the two. This type of barn is an adaptation to cold winter weather.**

PREVIOUS PAGES: *The barn in winter is a quiet and nearly mystical place.*

Barns for a New World

Early barns were built in one of two ways: by local custom or by plan. We sometimes forget that farming has been a scientific practice for centuries and that farmers have always spent considerable time and effort studying ways to improve their yield. The construction of a barn was a major investment, so any thoughtful farmer would carefully consider his design. Sometimes a farmer would independently develop a good barn design, but more often a farm journal offered a better concept.

During the late eighteenth century, most of the guidance concerning seeds, plants, and agricultural practices came to America from England. Many farmers subscribed to English journals that gave them practical advice on important agricultural issues. But the Revolutionary War ended the easy access to English seeds, tools, plantsmen, and information. Fortunately, the colonists had already developed a substantial amount of self-sufficiency by this time; now they were forced to organize their efforts to remain functionally independent.

During the days of the American Revolution there were few handbooks on scientific farming. Plantation owners like Washington who were interested in the latest theories would read about them in a paper of some sort. While there are few early books on barn construction, plans for improved farm buildings appear in early farm magazines, where there are dozens of examples.

RIGHT: **As farmers moved toward the western frontiers, barn designs began to change. In this instance, the door appears on the gabled end of the barn, now that threshing can be done in the field. The hay still goes overhead into the mow, but an adequate wind across the threshing floor is no longer an essential requirement.**

It is usually a surprise to realize that the oldest magazines in America are farm journals. Some are well over a century old, and back copies of historic agricultural publications are eagerly sought by collectors. One famous example known as the *Country Gentleman* had bloodlines dating to 1831, when it was called the *Gennessee Farmer*. It then became known as the *Cultivator* in 1839 and was merged with another weekly in 1866 to become the *Cultivator and Country Gentlemen*. It lasted well into the twentieth century and today its fans call it "The Country Gent."

Farm periodicals of all sorts covered rural America. Magazine historian Frank Luther Mott reports that, by 1890, 139 magazines were devoted to general farming. Add in all the specialized magazines—those devoted to dairy farming or poultry raising or threshing—and the numbers go much higher. These early periodicals, the oldest among them resembling a two-page newspaper, supplied the farmer with much of his information about farm innovation.

While literacy was not particularly widespread in early America, the notion that "scientific farming," as it was known, was a practice that could be taught was widely accepted. In 1855, the Michigan Agricultural College, the first college devoted to teaching farming, opened its doors. The college was preceded in its technical focus by the Rensselaer Institute in New York, the first engineering school in America. Training in farm mechanics at Rensselaer helped Jerome Increase Case develop his first mechanized thresher in 1842.

ABOVE: Milk cows need protection from harsh weather, but beef cattle and sheep are fairly hardy, and can spend their winters outside. The protected shelter under the barn is called a bye.

Along with education, land for agriculture was available to the new immigrant due to decisions made by the Continental Congress between 1780 and 1787 that helped push settlement toward the west. Later in 1862, President Lincoln signed three important bills that dramatically changed the face of American farming. The first was the Homestead Act, which gave each farmer 160 acres (64.8ha) of land if he was willing to occupy it for five years, the second was the Morrill Act, which established our system of land-grant colleges, schools that trained farmers all across America; the third was a bill establishing the United States Department of Agriculture. With eighteen divisions and hundreds of field stations, the USDA aided farmers through means such as collecting accurate weather information and supplying technical information on barn construction.

So although some might like to assume that the farmer of a century ago was an ignorant hayseed who built his barn the same way his neighbors built theirs, the fact of the matter is that most farmers built their barns and their agricultural businesses by using a fair level of technical expertise, rather than by resorting to local folklore.

BELOW: A connected barn, with buildings attached in a row, is a convenient arrangement. Beyond protection from the elements, it provides efficient use of space and materials, and affords farmers the opportunity to set schedules unaffected by outside influences.

LEFT: **Barns of the eastern United States frequently display the traditional New England saltbox-style roof lines that are typical of that region's architecture.**

The Ring Barn of Shelburne, Vermont

An overview of early barn structures in eastern America should include the Ring Barn in Shelburne, Vermont, an outstanding barn that, at the time, was the finest structure money could buy. While most barns are modest structures that fit quietly into the landscape, there are a few notable American barns designed by famous architects to serve the estates of the wealthy.

Noted for its size, its building date, and its outstanding quality of construction, the Ring Barn at Shelburne was designed by architect Robert Henderson Robertson, and was built in 1892. Designed for training and showing horses rather than for agricultural production, it has several unique aspects. The owner of the barn, Dr. William Seward Webb, wanted to introduce the hackney horse to America, so he requisitioned a barn design that was large enough to exercise and train them properly. At one time it was thought that this barn—at 418 feet (127.4m) in length—was the largest open span in North America.

The breeding, training, and sale of horses has long been the hobby of millionaires and the sport of kings. Dr. Webb, a railroad millionaire, and his wife, heiress Eliza Vanderbilt Webb, daughter of William Henry Vanderbilt, established a fabulous estate on six square miles (10sq km) of Lake Champlain Valley in the Adirondacks of Vermont. Their home included a manor house and three notable barns, one used as a dairy and another used exclusively for the training of hackneys.

Since the 1970s, Shelburne Farms has been operated by a nonprofit organization, serving as a pastoral preserve and environmental education center. Shelburne Farms also hosts special events like the Vermont Mozart Festival and houses a museum devoted to early American life.

THE INFLUENCE OF THE MENNONITES

Chapter Two

Buildings are tools.

— Gene Logsdon, *The Contrary Farmer*, 1993

Traveling across the Ohio Valley as far east as Philadelphia, one cannot help but notice the spectacular barns of the rural areas. Many of these barns were built a century or more ago, when barn building in America reached its peak. Constructed of virgin timbers, frequently hand-hewn and carefully fitted, these barns were built to last, and they certainly have.

In the late eighteenth century we begin to see a variation in the basic shape of the barn, as barn design began to feature an overhang, or forebay, an innovation brought to America by German-Swiss immigrants. These progressive barns, called Sweitzer or Swisser, were widely adapted

LEFT: A horse's dream come true, this elegant and spacious barn boasts a loafing yard and a beautiful place for "horseplay."

ABOVE: A substantial barn and silo built of glazed brick indicate the home of a prosperous farmer.

*BELOW: **Always an eye-catcher, a round barn attracts attention no matter where it is located. Round barns began to appear in the early nineteenth century, and most of the existing round barns are more than a century old.***

throughout the rural areas of America. In this case, the second level of the barn was larger than the lower level, providing an overhang to shelter animals below. It was a useful modification, designed to protect the barn doors and keep snow and wet weather at bay.

Of all of the barns built by the German immigrants who pioneered the area, none has left an impact that can be matched by the farm structures of the Shakers. The simple designs and beautiful proportions of their barns, granaries, and farm implements are still widely imitated by contemporary craftsmen.

The Round Barn of Lebanon

Perhaps the most notable example of a Shaker barn is the famous Round Barn of Lebanon, Pennsylvania. It's not entirely clear where the practice of building a round barn originated, but the most spectacular examples are associated with the religious group known as the Shaking Quakers. Their four-story round barn in Hancock, Massachusetts, was built in 1826, burned and rebuilt in 1869, and remains today as a masterpiece of American design. It is a model of efficiency, with stalls for fifty-two head of cattle around the rim of a giant circle on the ground floor, and a hay storage area about 55 feet (16.8m) in diameter and 35 feet (10.7m) high in the center. Hay was brought in by the wagonload on the top floor and dropped down to feed the cattle on the first floor. The resulting manure dropped to the bottom area below the stalls.

Round barns became quite popular during the latter half of the nineteenth century, when hundreds of round structures were built. Although the Round Barn in Lebanon is perhaps the most well-known example, round barns can be found throughout America. Today, more than a hundred examples of the round barn can be found in Indiana alone, a state noted for its efforts in preserving rural architecture.

ABOVE: **Prized for their elaborate interior framing, round barns are usually the pride of the local agricultural community.**

LEFT: **The great round barn in Pittsfield, Massachusetts, first built by the Shaker community in 1826, is perhaps the most famous barn in America. Damaged by fire and rebuilt in 1869, it is an important icon of American design.**

ROUND BARNS

The appearance of a round barn in our farmscape should be a reminder that one of America's most significant virtues exists in the right to religious freedom. Most of America's round barns were built by various utopian communities that consisted of people devoted to organizing a perfect society. While most round barns were built by religious groups, there were a few secular groups, such as the utopian community in New Harmony, Indiana, founded by socialist Robert Owen, who built them as well.

Historians say that the sharp rise in the formation of utopian communities came as a result of the Protestant Reformation. Utopian societies began to appear in America during the late eighteenth and early nineteenth centuries. They appeared in great numbers in the United States for the same reasons that other immigrants decided to settle here—namely because land was cheap, usually available without restriction, and religious toleration was a matter of public policy.

The Mennonites were the first significant group to build round barns. From their earliest appearance in America, these settlers became noted for their austere and beautifully proportioned architecture. Followers of Menno Simons, a sixteenth-century Anabaptist, the Mennonites spread widely throughout Europe and suffered various persecutions.

A group of English pietists, the Shakers, also left a strong legacy that continues to influence American culture and design. Arriving in America in 1774, a group of nine believers quickly captivated hundreds of others. Their first community was established in New Lebanon, New York, in 1787. Attracted to the communal life, many joined the congregation, and by 1826 there were eighteen Shaker villages in eight states.

In the early nineteenth century, round buildings of all types began to enjoy wide-

ranging popularity. Some point to a book written by Orson Squire Fowler of Fishkill, New York, as spreading the acceptance of the design. His book, *A Home for All; The Octagon Mode of Building*, is said to have popularized the style. Fowler was a phrenologist, a practitioner of the popular nineteenth-century study of interpreting personality and intelligence by analyzing the shape of the skull.

The round or polygonal style of building actually appeared in America at an earlier time. Thomas Jefferson, for example, is one of many who built an octagonal house. The style began to enjoy a substantial vogue again in the second half of the nineteenth century. That era witnessed an interest in designing buildings along "scientific" lines, and there was a great deal of experimentation with ways to improve domestic efficiency.

The construction of a round barn poses some significant design problems that are not present in conventional barn construction. Bringing all the structural members together and fitting them into a central point without sacrificing stability takes a master carpenter. For this reason alone of a round barn is a very special structure, one deserviing of our attention.

Round barns seem to fall into two categories: barns built as an expression of religious beliefs, and barns built in response to the scientific agricultural programs of the leading universities. Round barns seem to appear in nearly every state in the continental United States, but certain areas are more blessed than others. The award for the greatest number of round barns per square mile is hotly contested, with Wisconsin, Indiana, and Illinois all claiming the crown.

Indiana, a state noted for the preservation of its rural heritage, is reported to have more than one hundred round barns. Fulton County, Indiana, claims to have more round barns than any other county in America, and its town of Rochester hosts the Round Barn Festivals every year. In contrast, California has only eleven in the entire state, two of which are in Santa Rosa. But no matter what its location, age, or history, a round barn always attracts attention.

American farmers have always been interested in improving efficiency and productivity, and many built round barns because they had read articles about the benefits of this efficient design in the numerous farm journals of the day.

Building a Barn, Building a Business

Barns are America's oldest industrial buildings, the structures that allowed many hardworking and enterprising immigrants to become independent businesspeople. When the Homestead Act was signed in 1862, granting 160 acres (64.8ha) of land to any person willing to farm that land for five years, farmers suddenly had huge plots on which to toil. Managing a farmstead that large was tough for many pioneers, but improvements in farm technology made the job easier. Many of the barns on early homesteads were far more impressive than the homes of their owners, since the barn was the operating center for the economic well-being of a farm.

Choosing a good location for the barn took a certain degree of planning. For example, it was important to catch the right breezes for good ventilation. It was also important to be near a reliable source of clean water, and yet be situated so that there was some protection from storms and flooding.

A little handbook, first published in 1887 with the title *Farm Appliances: A Practical Manual*, reminded farmers to keep the comfort of the animals in mind. The barn can be a warm and cozy place.

*BELOW: **There are two schools of thought on hanging a horseshoe for luck. One says that the horseshoe should be hung with the opening on top so the luck does not "run out." The other says that any horseshoe— hung up, down, or otherwise—is lucky.***

A full hayloft provides wonderful insulation, and the body heat generated by a pair of draft horses, a cow, and a couple of calves raises the temperature in the barn noticeably. But the handbook notes that many farmers attempted to keep the heat in and did not provide enough ventilation to keep their livestock healthy. "Warmth alone is not comfort," *Farm Appliances* lectures. "An animal may suffer from cold in close, damp impure air, which is really warm, while it will be quite comfortable in fresh pure air which is below the freezing temperature."

The handbook goes on to provide guidance on how to provide adequate ventilation without letting in chilly drafts, and how to ensure adequate light to do farm chores and still make the barn tight and comfortable for the farmer and his animals. Many farmers already understood the importance of these basic guidelines, and the solid barns they built remain as testimonials to their wisdom.

*LEFT: **The nineteenth century was the heyday of American barn construction and, for a brief period, barn buildings echoed popular residential architecture of the time. The decorative windows on this barn are typical of American Gothic architecture.***

*BELOW: **The rise in the popularity of English-style architecture marked a significant change in barn construction, as doors were now placed at the ends of the barn rather than on the side. This placement choice indicates that threshing was no longer the primary design consideration in barn building, as it could now be done more efficiently in the field by machines.***

English, Dutch, Deutsch, Der English

If someone asked a fairly knowledgeable barn enthusiast, "What do you consider to be the typical early American barn?" the answer would probably be, "Oh, it's the English barn." In fact, early barns in some regions were more closely modeled after the architectural designs of the Dutch. The typical shape of the common American barn began to evolve in the mid- to late eighteenth century, but it did not necessarily resemble the barns of England. The first barn designs from England, imitated by our early colonists, were long, low buildings, topped by a thatched roof, while the "English" barn has a large shed shape.

The English barn came into widespread use in the eighteenth century and was so common that it was referred to as the Yankee barn or the three-bay barn. When thoughtfully built, it is a structure of satisfying and elegant proportions. Generally two stories in height, it has large central doors able to accommodate a hay wagon through openings in the center of each side. The hay is stored in the loft, located on the upper floor. The large open area in the center of the barn is the threshing floor, used to thresh and

winnow the grain after harvest. While most English barns had some venting to provide air flow, the barns were generally without windows. The English barn flourished in that time and place in American development when wheat dominated the economy.

In general, a home for the livestock was not an important consideration in early barn construction. Most animals were slaughtered in the fall so they did not have to be fed through the winter. Horses were not in common use as draft animals until the late eighteenth century and, not until the development of the silo in the 1880s, would a year-round beef and dairy industry develop.

Historian Eric Sloane tells us that the true Dutch barn exhibits more European design elements than any other type of barn built in America. He also notes that there are no such barns in the Netherlands, though the early Dutch settlers in New York were known for building barns with tall, peaked roofs. In these barn's the doors appear on the ends of the barn, rather than the sides. Sloane surmises that this style of barn was adapted from a variety of European sources.

The Dutch barn, also known as the New World Dutch barn, has sloping rooflines and low side walls, a distinctive shape that makes it an easily identifiable rural structure. The Dutch Barn Preservation Society notes that this unique style appears in rural New York as well as northern and central New Jersey. Built by the early colonists who emigrated from the Netherlands and settled in the Hudson Valley, these barns appeared in large numbers for nearly two hundred years, from about 1630 until around 1825. The longevity of their design is a direct result of their usefulness.

Knowing the severity of the New England winters, it is easy to surmise that the roof shape the Dutch settlers used developed in response to the heavy snowfalls in the area. The framing of the Dutch barn is also unique, since the weight of the tall rafters does not rest on the exterior walls. Instead, framing is based on a system of internal support beams. One side of the wagon doors may be divided, another unique characteristic. Dutch barn builders encouraged the habitation of swallows, so many barns have small holes in the gable designed to attract these insect eaters.

Many of the most spectacular barns in eastern America, large three- and four-story structures of beautiful craftsmanship, were built by the Pennsylvania Dutch. As any knowing native will quickly explain, the Pennsylvania Dutch are actually emigrants from Germany (Deutschland), not Holland.

The Pennsylvania Deutsch, or Dutch, as they are commonly called, immigrated to the Philadelphia area in 1683. Followers of Menno Simons, they came to America to avoid religious persecution. The Mennonites, as they are known, became pioneers in Pennsylvania, Virginia, Ohio, and Indiana, and developed a style of barn construction to suit their agricultural needs.

PREVIOUS PAGES: **In addition to the fashion of building round barns, the early nineteenth century saw a flurry of extensive construction focused on building extremely large barns in the New England area. Nearly five stories tall, this a particularly fine example.**

ABOVE: **Natural flycatchers, a family of barn swifts are a welcome addition to the farmyard for the service they provide.**

OPPOSITE: **Two important farm tools are shown here: the tall silo, which first made its appearance in 1873, and the mechanized manure spreader, a wagon that allowed the manure to be distributed evenly across the surface of the soil.**

RIGHT: **Because the Amish do not use electricity, horsepower still provides much of the muscle on the farm. For this reason, a comfortable barn to house horses and their equipment is a necessity.**

BELOW: **It's difficult to imagine the effort it takes to maintain an Amish farm, but operations like these are actually experiencing a growth period, perhaps due to disenchantment some people feel living in a fast-paced, technology-based society.**

Mennonite families farm using horses as a source of power, so their barns are designed to accommodate several large, heavy draft animals. Many Mennonite barns are situated on the side of a hill or bank, which allows the upper floors to be used for hay storage and the bottom floor to be used as a stable for the horses. Hay and feed can be thrown down easily from the mow on the upper levels into the feed boxes in the stalls below.

And a final and confusing word about differentiating the Dutch and the English, in barns and in local custom. To a Mennonite or other member of the Old Order Amish Church, any outsider, in other words, a person who is not Mennonite, is called "English."

Jacob and Anna Yoder and Their Barn

Holmes County, in northeastern Ohio, has a large population of Mennonite farmers known as the Old Order Amish. Jacob and Anna Yoder, along with their children and grandchildren, operate their farm in an area that has been a Mennonite stronghold for about two hundred years.

Draft horses provide most of the power needed to operate this farm, in the traditional Amish manner. Five blond and beautiful Belgian draft horses occupy the barn on the Yoder farm and are an integral part of the farm operation. On this Holmes County farm, the draft horses occupy stalls on the lower level of the barn, next to their wagons and other equipment. A buggy horse is stabled there too. Handsome thoroughbreds too slow for the racetrack frequently show up as Amish buggy horses.

*ABOVE: **The windmill provides the power to pump the water necessary to supply the farm. Once commonplace across America, the development of municipal water supplies has relegated the windmill to the more isolated farmsteads.***

Comparatively speaking, the Yoder farm is a small operation, with only 80 acres (32.4ha) of land. The economic base for the Yoder farm is veal, raised under kosher practices, some of which is shipped to Israel. The Yoders also raise their own oats and hay to feed their horses.

The Yoder barn and related farmyard buildings are examples of essential industrial buildings on the farm. Their barn is a banked barn, with the horses, the hitches, and the harnesses on the lower floor. The upper floor is used for grain storage. While the main house is a traditional farmhouse of approximately the same vintage as the barn, a

*ABOVE: **It's easy to identify an Amish farmstead. All you have to do is look for beautifully maintained buildings without any signs of power poles or telephone wires.***

small nearby "doodie" house (a house for the grandparents), is a contemporary structure with premanufactured metal siding.

Following traditional Amish practice, the farm has no electricity and the family uses kerosene lamps for light. Power to operate a small thresher, binder, and hay baler is provided by a small diesel engine, which is pulled by the horses. Using a machine to power the thresher mechanism relieves the horses from having to perform the dual task of powering the binder as well as pulling it. After much discussion and consideration, some modern innovations have been allowed in some traditional Amish congregations.

Barns on Fire

One of the most interesting features of an old barn is its lightning rods. Especially rare are lightning rods with a glass ball, which frequently turns lavender with age. Reading one of these glass balls is about as easy as it gets—if the glass is shattered, then lightning has struck. Because the barn is often the tallest structure in open rural areas, it is attractive to lightning, which is responsible for nearly one-fourth of all barn fires.

*OPPOSITE: **As the price of farmland increases in traditional Amish strongholds such as Holmes County, Ohio, and Pennsylvania, Amish families have moved west. This farmstead appears near Jamesport, Missouri.***

In his book *An Age of Barns*, barn illustrator and meteorologist Eric Sloane notes that some farmers refused to use lightning rods due to personal beliefs that lightning was evidence of God's will. And Eric refuses to seek the shelter of a barn in a thunderstorm, because the static electricity built up in a barn by drying hay attracts lightning. It's a curious phenomenon, one now nearly forgotten by our predominantly urban population.

Storms are only part of the fire problem, though; some barn fires are caused by spontaneous combustion. Hay and other forage crops with a moisture content over thirty percent are potential fire hazards. If the material cools too slowly, it may turn brown or moldy. And if the heat cannot escape and air gets in, a spark may spontaneously ignite.

Even silos have been known to burn. Although fodder stored in a silo is supposed to have a high moisture content that makes it less susceptible to fire, if the fodder drops below forty percent moisture content, it can ignite. Over the years, farmers have developed some practical precautions against this danger: Making sure the grain and the hay are dry and well cured before storing them in the barn helps reduce loss. Placing hay in several locations or even resorting to the traditional haystack are other solutions to this problem.

FOLLOWING PAGES:
Spring plowing begins with a pair of draft horses. The modern American economy was built by farmers who were able to become independent businessmen.

Building a Barn: The Amish Tradition

Building a barn is easy: just invite the neighbors over and fix them dinner. Well, that's a tremendous simplification, but there are some fascinating traditions that characterize Amish and Old Order Mennonite barn building. If the barn under construction is a replacement barn (one meant to replace a barn lost to fire or other disaster), custom requires that the structure be rebuilt within thirty days. (While the Amish and the Mennonites share basic beliefs and, historically, many traditions, the Amish and Old Order Mennonites follow the dictates of "plain living" more strictly than modern Mennonites.)

For the Amish, barn building is a community project in which everyone in the area pitches in to help. Because of the tremendous organization and planning that goes into the preparation for construction, an Amish barn raising takes only one day. When three or four hundred men show up to work, it represents a tremendous outlay of time and labor. Designing the project so that it takes only one day allows the neighbors to return home to tend to their own daily chores and feed their livestock.

The barn building project is designed and coordinated by a member of the community who is an experienced barn builder and specializes in the craft. While the Amish practice the simple life and use traditional farm practices, they are innovative in many respects. Amish barns built today use standard 4-

BELOW: **We usually associate traditional barns with Mennonite barn builders, so it sometimes surprises observers to see a barn built using contemporary materials. Erecting a barn in a day is the Mennonite way; in this matter, they are masters of prefabrication, sub-assembly, and teamwork.**

OPPOSITE: **Simplicity is a characteristic of the Amish farmstead. Good barns, carefully maintained, will last several lifetimes and can be passed along to future generations.**

by 8-foot (1.2m by 2.4m) sheets of plywood and other contemporary materials. Foundations can be formed of precast concrete blocks rather than hand-hewn stones.

Even so, building an Amish barn is a special occupation and only a precious few are experienced in the trade. Like any other seasoned specialized contractor, a good Amish builder works by referral and is never short on work. When a barn needs to be built, the call can go out across communities in several states, searching for a builder who has the time available to coordinate the project.

Another interesting aspect of Amish barns is that the barn building is a community project because the barn is usually financed through Amish institutions. Having the barn built by local community members is another way of insuring their investment.

The last significant feature of an Amish barn raising is food. It takes almost as much preparation and coordination to feed several hundred workers and their families as it does to prepare for construction. The wives of the carpenters spend their day cooking and serving countless meals, as well as keeping the crews supplied with gallons of coffee, lemonade, and other vital refreshments. When the Amish build a barn, all the men, women, and children—the entire community—are called upon to help.

FOLLOWING PAGES:
Barn construction is a community activity, but all the preliminary work on the design and selection of materials is handled by an expert recognized in the Mennonite community for his skill. It's a very specialized craft and the master barn builders are known throughout the Mennonite community and booked months ahead, some traveling across the country when their unique talent is needed.

BARNS
OF THE
MIDWEST

When I close my eyes and picture an American barn, I see a red barn with a gambrel roof. My barn is painted red with white trim. There is a silo. The barn of my dreams is in Iowa. It's an agricultural archetype, where Old McDonald had his farm and became an early icon for the American way of life.

Farms across the middle-American states were built during a period after the Civil War when the territories were expanding, immigration was encouraged, and scientific farm practice was eagerly embraced. The Homestead Act of 1862 made agricultural land available, and railroads were opening new territories and taking products to market. Steam power and mechanization came to the farm, providing some important tools and greater efficiency.

LEFT: **Chopped corn and other ingredients fill the silo, providing a nutritious winter diet for cattle. The development of the silo and the right mixture of feed meant that Americans could eat meat all year round. Before the silo, it was prohibitively expensive to support cattle through the winter months.**

ABOVE: **A double gable with board and batten siding indicate that this is a barn from the late nineteenth century, when farmers enjoyed including elements of Victorian Gothic style in their barns.**

The second half of the nineteenth century saw tremendous changes in the design of farms, driven by a number of factors: Farmers began to specialize, mechanical equipment was made available, and urban markets became more accessible. In addition, the land grant agricultural colleges in such small towns as Ames, Iowa, and Racine, Wisconsin, would eventually become world famous for turning out hundreds of knowledgeable farmers, the industrial entrepreneurs of the day.

Improved barn design was only one of the many curriculum areas that farm students could now address. By the late 1800's, there was a clear awareness that farming was a business rather than simply an occupation, that hired help was scarce and mechanization could improve production, and that the industrious farmer was a highly independent executive. The railroads, the telegraph, and the improved postal system kept farmers aware of their markets. The creation of the United States Department of Agriculture meant there was a bureau to assemble statistics. Farmers now had reliable historic data, including weather information, soil analysis, and hundreds of other important items of interest. All these changes would eventually have an impact on farm practices and barn design.

ABOVE: **The gambrel roof is an efficient way to attain a maximum amount of storage space within the rafters. The gambrel roof on the ventilator is a charming detail, providing a sort of visual pun for the observer.**

Looking over the Landscape

In general, the barns of the Midwest are closely related to the kinds of crops grown in the area. The barns that cover the Great Plains of Middle America, reaching from Canada to Texas, vary widely in shape and style. But while eastern colonists built barns shaped by the need for threshing space and designed around lines of ethnic heritage, the barns of the central American states were designed to support a specific group of crops.

Barns with a gambrel roof are very common in the Midwest, so common that we refer to this distinctive shape as a "barn roof." This rafter design solution provides a lot more usable space on the upper floor, and is sometimes used for houses. Even then, we refer to it as a barn roof.

In addition to the introduction of the gambrel roof, a second important architectural addition began to appear on farms across America in the late nineteenth century. Historian Wayne D. Rasmussen attributes the first upright wooden silo—that round tower frequently located adjacent to the barn—to Fred Hatch of McHenry County, Illinois, in 1873. Silo construction began in the 1870s and spread both east and west. It quickly became one of the most important special-purpose structures on the farm, as common as the granary or corn crib.

*FOLLOWING PAGES: **An ice storm in late spring can be as devastating as it is beautiful. Ice and hailstorms are of short duration, but they can kill crops, break tree limbs, and bankrupt a farmer.***

*BELOW: **The gambrel roof, which we now commonly call a "barn roof," comes in several shapes and sizes, depending on the whim and custom of the builders. After about 1880, plans for framing could be acquired from farming magazines and local Farm Bureaus; these advancements in information-sharing had an enormous influence on the roofing styles of the day.***

Silos

In general, a silo is any structure used to contain green chopped fodder for livestock. Farmers back to the time of the ancient Romans preserved green feed by storing it in underground pits, keeping light and air away from the material in order to preserve freshness. But silage began to come into its own in the second half of the nineteenth century, due in great part to the serious study of crops in the new agricultural colleges of the Midwest.

A certain amount of acid must be present in the plants used for silage in order to preserve the feed for the livestock. The critical acids are lactic acid (as in sour milk) and acetic acid (as in vinegar). As plants begin to break down, they can produce these acids, which tend to preserve the plants in a way somewhat similar to the way vinegar preserves pickles. Since clover, alfalfa, and soybeans have little sugar or starch, they cannot produce these acids and do not preserve well in a silo. Good crops for silage include crops with high sugar contents like corn, sorghum, and milo. These crops can be combined with clover and alfalfa to create a nutritional feed—a sort of coleslaw for cows. It takes the right combination of ingredients to make the taste just right and get the acids to balance, but cows love it as a winter feed.

*BELOW: **Silos were also built by contractors who began to specialize in the construction of reinforced concrete towers. The weight of the damp silage puts a tremendous load on the walls of the tower, so careful and experienced design is required to keep the tower from collapsing.***

The tall, distinctive appearance of the silo, which seems to be a universal part of many farmyards, provides some significant information to those who are able to read the story. The appearance of a silo on a farm indicates that this is a mixed farm that feeds livestock, and that the farming operation was probably started within the last century or so.

The use of silos spread east, and most of the silos seen in the eastern states were built after the 1873 introduction date. The weight of silage puts a tremendous pressure on the interior walls of the silo, and a round structure distributes the weight most evenly. Silos are usually built with steel bands to reinforce the walls. This may explain why so few silos were built of stone, since it would be difficult to build a round stone structure with perfect load distribution. Still, plenty of stone silos can be found in the New England area.

ABOVE: **Where practical, a stone silo is a sturdy and useful structure. This one is not very tall, but the use of the native stone which matches the foundation of the adjoining barn makes for a very handsome presentation.**

Reinforced concrete (concrete in which metal has been embedded for added support) began to appear in silos and other farm buildings well before 1900, again due to the engineering curricula in the agricultural colleges. The use of lime and gypsum as a cement had been well known to Roman engineers and builders, but portland cement (a heated mixture of ground limestone and clay) wasn't invented until 1824, when it quickly became the dominant cementing material.

The development of reinforced concrete is attributed to a French gardener named Joseph Monier, who patented his breakthrough in 1855. Practical applications for reinforced concrete began to appear in America during the 1880s, and by 1900 the use of tie rods and other structural steel was in common use. Thus, concrete silos began to make their appearance at about this time.

The development of the silo meant that fresh meat became available to the American consumer year-round. Before the silo, most farmers had difficulty keeping their cattle during the winter months. Slaughtering was done in the fall, and meats were commonly cured or smoked to keep them through the winter. Providing enough feed to keep a dairy herd through the winter was a tremendous challenge. The advent of the silo, combined with the development of railroad refrigerator cars, now brought a supply of inexpensive fresh meat to the cities, making the American diet considerably less dependent upon seasonal variations.

The Lotenvitz Barn of Charles City, Iowa

There are hundreds of barns across middle America and many of them still do daily duty as the heart of the farming operation. Charles City, Iowa, is perhaps the stereotypical American farm town. The Lotenvitz barn, located on the outskirts of Charles City, seems to typify the classic barns of the rural Iowa countryside.

Charles City is historically noted for being the birthplace of the American farm tractor: the Hart-Parr tractor was developed here in 1902. The city is also noted as the birthplace of the American Beauty rose, that long-stemmed hybrid popularized in the 1890s. And it is home to Salisbury Laboratories, a farm industry that started out developing specialized chicken feeds, and wound up helping thousands of American farm wives transform their "egg money" into substantial income. These new feeds enabled large flocks to remain healthy, and raising chickens suddenly went from being a backyard industry to a major economic activity.

Owned by Ivan and June Lotenvitz, the Lotenvitz farm is located on a rural route amid Iowa corn-fields. The Lotenvitzes have long since retired but still live in the farmhouse, where Ivan was born. Fields and pastures are easily leased to others for crop production.

*BELOW: **A well-built barn can be a community asset for generations. This dairy barn is now part of a local historic site.***

The tidy farmyard contains all the usual farm structures, including the stereotypical Iowa barn. The barn was built in 1926, a replacement for an earlier structure. Its gambrel roof provides an efficient way to get the maximum amount of storage space for fodder under the rafters. The first floor is built of brick, not uncommon in this part of northern Iowa, where winter weather can be severe.

The Lotenvitz barn has a slight overhang just under the eaves where the hay hook is secured. Loading fodder into the upper loft was accomplished by fastening the hay bales onto a hook and hoisting them into a loft. It was only slightly less tiring and time-consuming than the earlier method of tossing hay with a pitchfork.

The Lotenvitz barn typifies the twentieth-century barn. Built on a concrete foundation, it is the most popular design for a barn, combining modern materials with a traditional architectural plan.

*BELOW: **Hay was once packed into tidy square bales, convenient for stacking and storing. New machinery and techniques now allow the hay to be quickly rolled instead.***

Painting the Barn

To the purist, there are only two colors to paint a traditional barn: whitewash white or barn red. White barns and other farm buildings were painted annually with a mixture of casein (derived from skim milk), lime, and a little linseed oil. Spread liberally over the chicken coops, barns, and fences, the paint provided reasonable protection from the effects of weathering, but it was not very durable. It didn't matter, though—it was cheap.

Red is the traditional color associated with barns. The red shade was originally created by adding iron oxide, a comparatively inexpensive pigment, to a mixture of linseed oil and turpentine. The resulting color did not show dirt readily, but its primary virtue was weather protection.

Barns were not the only structures for which barn red was the preferred color. Since it was relatively inexpensive and held up well, it was used for many, many industrial buildings. The color was universally used by the early railroads and, in some circles, this dark, earthy brick color is known as boxcar red or caboose red.

ABOVE: **The rolled hay bales shed water as effectively as the old haystacks used to. It's a much more efficient way to harvest and store feed.**

FOLLOWING PAGES: **Barns on the western plains rarely have a silo; the cattle make do on the existing range and any supplemental hay is trucked to them.**

71

RIGHT: **Eastern Texas features a unique style of barn, one made of logs that are each about twelve feet (3.7m) long. Here, the original log barn at the Johnson Ranch has had a stone addition and a more conventional barn roof put over the earlier construction.**

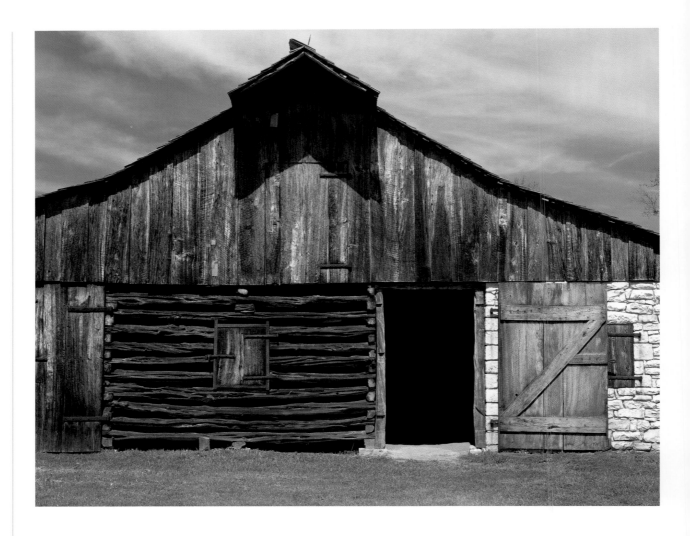

A freshly painted barn was once the hallmark of an industrious farmer. Indeed, a person's local reputation rested on the condition in which he kept his barn. Midwesterners who are used to nicely painted barns are usually contemptuous of the unpainted redwood barns of California and the Pacific Northwest. It's tough to convince them that the silvery gray weathering is the proper color for a California barn. To an easterner, the unpainted barn is just another indication of the creeping moral decay of the Californian.

Log Barns of Eastern Texas

To most folks, Texas is still the Wild West. Even today, urban residents of Texas towns can be seen wearing boots and big hats, recalling cattle drives and cowboys. Texans themselves do not seem to think of themselves as part of the Midwest; to them, Texas is still a world of its own. For purposes of vintage barn construction, though, Texas has to acknowledge a few ties with eastern America.

One of the most interesting types of barns in eastern Texas is the log barn, a structure made from hand-hewn logs. In general, the log barn is a little smaller than the great barns of the eastern United States, where tall trees, building stone, brick kilns, and water-powered sawmills provided other building options. Log barns appear in many rural regions, but are especially prevalent in the Appalachian area, eastern Texas, and in the high prairies of Colorado, Montana, and Wyoming. Log barns have been documented in eastern Texas as early as 1840.

Small and without lofts, log barns are composed of two to four comparatively small structures, about 12 feet (3.7m) square. Terry G. Jordan, who has done considerable research on the Texas log barn, describes them as "carelessly built" of unhewn timbers and saddle-notched corners. This type of construction left plenty of opportunity for ample ventilation.

Texas log barns typically have two little buildings set about 10 feet (3m) apart, providing a central passageway between the two sections. A single roof covers all three areas, making the entire enclave about 14 feet (4.3m) wide and perhaps 35 to 40 feet (3.4 to 3.7m) in length.

Log barns seem to have been used as corn cribs or granaries and perhaps for a small animal or two. The weather in Texas is warm enough to leave animals outdoors all year, and barns seem to have been a precaution primarily against theft, rather than for weather protection. Like barns in general, though, log barns are rapidly disappearing. They seem to be more vulnerable to destruction than vernacular structures in other parts of the country, since their usefulness as functional shelters and importance to regional culture are not always recognized.

*BELOW: **Barns and sheds utilizing entire logs in their construction are common in the higher elevations of Colorado, Wyoming, Utah, and other states with great forests of lodgepole pine. Log barns can survive for decades with little maintenance.***

Frank Lloyd Wright's Midway Barn

Midwestern barns come in all shapes and sizes, from the modest vernacular structure to the grand, architect-designed estate. Known for his innovative approach to architecture, Frank Lloyd Wright designed buildings that were always unconventional. He is noted for developing what he called "organic architecture," which aims to design buildings that harmonize with their environment and their occupants. His unorthodox Midway Barn is usually omitted from books discussing his designs, although it is just as compelling as his other creations.

Wright was born in Wisconsin in 1867, just after the Civil War. He studied engineering in Wisconsin and then went to work in Chicago, first working under Louis Sullivan, then developing his own private practice. By the turn of the century he was already noted for his extraordinary "prairie style" houses and for his unconventional commercial buildings.

Always a controversial figure, Wright's private life was as unsettled as his professional career. He fell in love with Mamah Cheney, the wife of a client, and a scandal ensued. He then left his own wife and

*BELOW: **The Midway Barn at Taliesen was designed to fit into the landscape rather than to sit on top of it. It is a marvel of Frank Lloyd Wright's architectural mastery, and exemplifies the diverse nature of his vision.***

family and retreated to the country, turning an estate that belonged to his mother into a private home for himself and Mamah. The innovative complex included a design studio and workshop, along with accommodations for apprentices and other staff. Wright named the complex Taliesen.

Designed to be self-sufficient, Taliesen (Welsh for "shining brow") boasted a working farm that supplied the compound and its citizens with food. In this manner, students who came to Wright's studio to study as apprentices usually found themselves working on the farm as well.

In 1914, a tragic fire, possibly set by an underpaid and jealous worker, killed Mamah Cheney, her two children, and four employees of Wright's staff. The complex was rebuilt and expanded after the fire, and Wright continued to acquire acreage to add to the farm, though he was devastated by the disastrous event. As his workload grew, and the number of his students and staff members increased, additional farm buildings were erected to provide and store food for everyone in the complex.

ABOVE: *Weather vanes can be quite decorative, but they perform important functions as well. The vane provides critical information about wind direction to the concerned farmer attempting to protect his crop. Likewise, lightning rods are more than mere decoration. A lightning rod can be an important safety device during storms, as the barn roof is frequently the highest point in the landscape.*

Wright designed the Midway Barn for his estate in the late 1930s. Like his other designs, this barn incorporated many innovations, from grand-scale changes concerning the layout of the silos, to minor aesthetic issues involving the placement of the weather vane. He was nearly seventy years old when he designed the Midway Barn, and over time the ideas he had developed about certain farm practices had significantly influenced his design concepts.

Since he had lost three hay barns on the property to fire, Wright refused to store hay in a barn any longer. Instead, hay was left outdoors in stacks. Wright wanted a progressive dairy operation on the farm, but chose to design his dairy barn according to his own principles rather than standard dairy practices. His cows developed mastitis—and reportedly never did very well—most likely because the hay, stored in the open, was substandard.

Later, Wright designed two short silos and placed them across the road from the barn so that their shape would not interfere with the view of the farm. Most farmers have the silo adjacent to the barn so that the feed is close to the animals. But even years of experience with harsh Wisconsin winters did not convince Wright to place the silos in a more convenient location.

The Midway Barn is a long, low structure that follows the side of a hill with no tall silos to obstruct the horizontal lines. Painted a traditional red, the barn is surrounded by fenced pastures where the cows and calves can graze.

THRESHING: THE HEART OF THE BARN

EARLY BARNS IN AMERICA reflect the nature of the crops that were grown at the time. Most of our early barns, particularly in the eastern half of America, are centered around a "threshing floor," since grain was our most important food crop. This floor is the heart of the barn, the area where the years' crop was processed and stored.

The threshing floor is a large, open, central space with a tall ceiling. In this area the sheaves or shocks of grain that had been stored in the mow were pitched and beaten with flails to separate the heads of grain from the stalks. Sheaves were loaded onto carts and then driven into the barn, where they were stacked on one side until it was time for threshing. Separated grain was stored in a secure but ventilated granary, safe from theft and vermin. Straw was pitched into the loft.

Early threshing, usually pronounced "thrashing," was a time-consuming and labor-intensive process. Today we commonly differentiate among crops when we talk about the various types of grain we grow. We describe them as wheat, barley, rye, or buckwheat. But our agricultural ancestors from England referred to all these grasses, with their heads of closely packed kernels, as corn.

When we think of corn, we are usually describing a grain that comes on a cob rather than a stalk. The native Americans introduced the early colonists to this plant. Maize, as it was called, was neither popular nor widely used as a staple in early colonial times, and was looked down on as being inferior to the other grains.

An enormous amount of labor was required to process the crop in the field and turn it into something suitable for human consumption. Farmers needed to use both the stalk and the head of grain for food and they needed to harvest at just the right time of the season. Dorothy Hartley, a writer who chronicles the agricultural folklore of medieval times in her book *Lost Country Life*, tells us, "When the barley hangs its head and the oats begin to shed and wheat stands stiff and begins to open," it is the time for harvest. After being cut, the grain is bundled into sheaves and left to dry in the field. The length of time needed to dry the sheaves varies with the type of grain and, of course, the weather.

Hay has to be cut when it is green and full of sap, but must be harvested just when it starts to dry. Damp hay that has been closely packed can heat until spontaneous combustion occurs, so many small haystacks are better than one large one.

Sheaves of grain were transported to the barn and stored on either side of the threshing floor. Barns were built facing the wind and sun, so that a strong draft could enter through the doors and sweep across the threshing floor. Good light was also important to threshing work, since separating grains takes keen eyes. Bundles of grain were laid in the middle of the threshing floor and the threshers stood around in a circle, using flails to beat the grain and separate the head from the stalk. It was tiring and time-consuming work.

After separating, the heads of grain would be gathered up and tossed, removing the chaff or husks by fanning or winnowing. There were several ways to do this, such as by tossing basketfuls of grain into the air and letting the breeze separate the grain from the lighter chaff, or by using a shovel and throwing the grain onto a pile, letting the wind blow through the falling kernels. The best kernels of grain were then selected as seed for the next year's crop, a chore for small, agile fingers and sharp eyes.

When mechanical reapers and threshers appeared in the nineteenth century, there was no longer a need for the traditional threshing floor. The central American states were being opened for development during a time when mechanized harvesting equipment began to appear, so traditional threshing floors are rare in the Midwest. As we've seen, barn architecture almost always reflects the important changes in the technology of the times, and thus the threshing floor was slowly phased out of new barn designs.

Harvesting took only a few days, but threshing was an activity that went on for months at a time. Bundles of grain were stored in the barn and beaten, and one by one the tiny kernels were separated from the stalks. It was a labor-intensive process, made much easier when J.I. Case developed a mechanical thresher.

The Dairy Barn

Most farmers kept a cow or two to supply milk and butter for the family. Even in colonial times, cows were common, and were sometimes used as draft animals. The practice of keeping cows solely for the production of milk dates back to the days of the pharaohs, but the dairy industry itself is a twentieth-century development.

Dairy barns tend to be of more substantial construction and better maintained than other barns, and for good reason. The dairy barn is in heavy use every day of the year and must meet high standards of cleanliness, so it is designed with those two considerations in mind. While other barns leave spaces in the siding to encourage ventilation, dairy barns are solidly built. Good air circulation helps the hay in the loft to dry, but dairy cows need to be protected from cold drafts, and the milk needs to be kept free from dust and other debris.

BELOW: **Windows all along the lower level of the barn indicate that this is a dairy barn. Milking is an activity that requires ample amounts of good lighting.**

Dairies are common across America and indeed are found in every state, but the upper Midwest is the region we usually associate with processing milk products. Fresh milk, cream, butter, and cheese come from Wisconsin and the other states around the Great Lakes. The development of the dairy industry really began during the second half of the nineteenth century, at about the same time that homestead lands in the Midwest were opened.

The practice of improving the performance of dairy cattle by selective breeding and by keeping detailed production records began in England in the middle of the nineteenth century. Pasteurization of milk, making it very safe to drink, began around 1895. Early farmers who kept dairy cows needed barns located near a spring or reliable year-round water source, since milk needed to be cooled immediately after it was produced in order for it to keep. But pasteurization helped ensure milk's purity and allowed the farmer some latitude in choosing the location of a dairy.

Milking machines were not introduced until the early part of the twentieth century and were not in widespread use until after World War II, when rural electrification helped make milking machines a practical solution for small, family dairy operations. Because cows have to be milked in the morning and

ABOVE: ***Dairy barns are generally built of heavier materials than other barns. Because dairy barns had to withstand a lot of heavy use on a daily basis, a more substantial construction was needed.***

81

at night, a dairy farm was an extremely labor-intensive operation, one that required either a large family or hired hands in order to manage it properly. The size of a dairy herd was therefore limited by the farm's milking capabilities.

Dairy cows are sensitive animals; some might say temperamental. They are creatures of habit and tend to get very nervous when their environment is not to their liking. A dairy barn must be clean, light, just the right temperature, and free of distractions in order for cows to produce milk.

Dairy barns usually have concrete floors that make it easier to maintain sanitary conditions. But concrete is cold and not the best surface for delicate dairy cattle during the winter. Because of this, dairy farmers have adopted a number of strategies to ensure the comfort and productivity of their cows, including using the wooden flooring that may still remain in older dairy barns.

Dairy barns usually have a number of related structures close by. The silo is most often adjacent to the barn, as are several structures for processing and cooling the milk. Milk production has always been one of the most profitable of all the farm industries, but it also requires a considerable initial investment. For example, a good, productive herd of Holstein or Jersey dairy cattle is much more expensive to acquire than beef cattle. In addition, the investment in a milking parlor itself requires a substantial outlay of funds, since every parlor needs pumps and milk coolers, along with the integrated electrical system required to manage such a complex operation.

Poking Around the Barn

Throughout most of American history, the farmer was an independent businessman. Evidence of his ingenuity and inventiveness is frequently reflected in the various old barns around the country. Every farmer had a few tricks to make his life a little easier, and farmers were always experimenting with improvements in stalls, feed boxes, and watering troughs. Techniques for dropping hay and feed from the upper lofts to the animals below were subject to constant experiment. New and improved methods for cleaning the barn and storing the manure to proper age were continually being investigated.

Mechanized equipment such as mowers and reapers saw widespread use after the Civil War. Most farmers could make small repairs themselves and many had a profitable business in blacksmithing. There were always plows to sharpen, horses that needed shoes, and mowers with missing teeth. And in the days before widespread electrical power, a farmer needed to be capable of lifting heavy bales to his hayloft. He needed to pump water, pull stumps, build fencing, and do dozens of other chores related to farm production. Proof of his inventiveness and the tools he used can be found in the barn. An efficient farmer was a saver of string and baling wire, of old coffee cans full of bolts, nuts, and washers. Old pumps, worn-out gas engines, used tires, leftover shingles, and an extra brick or two were often stashed in the barn or workshop. Barns were also used for storage, and all sorts of wonderful treasures sometimes appear from the corners. Old furniture, outgrown bicycles, and that painting that Mama hated all find their way into the barn. Every barn has stories to tell; some are written by its architecture and some are written by its occupants.

*BELOW: **Barns with simple shed roofs are common throughout the West, taking the place of barns with the gambrel-style roofs that are essential for hay storage. The absence of both a silo and the large hay loft indicate that there is less need for hay storage, since animals can forage outdoors in the milder Western climate.***

BARNS OF THE WEST

The barns of the West offer a distinct contrast to the farmyards of the Midwest and the eastern seaboard. It's difficult to believe that many of the simple wooden barns of California, for example, are of the same vintage as the magnificent, carefully crafted stone structures found from Pennsylvania to New York, but in truth, many of them are about the same age.

Visitors from east of the Rockies sometimes say that Californians need to paint their barns, commenting that western barns look old and decrepit with all that gray, weathered siding. Visitors are apt to poke fun at those fun-in-the-sun Californians, indulging in white wine and hot tubs instead of painting their barns like the hardworking midwesterner. But California barns, built of old-growth redwood and Douglas fir, need no paint and are best left untreated.

The weather is different west of the Rockies and the barns there reflect that difference. Generalizing about barns in such diverse western terrain as

LEFT: Built to support the cowboys, this horse barn provides some fodder and some protection from bad weather for their hard-working horses. Many of these western barns are only used a few weeks of the year.

ABOVE: Hay barns, such as this one, frequently stand alone in the prairie. They are built on the range for the sole purpose of storing hay, and are not a part of daily farm operations.

ABOVE: **Buffalo and range cattle can survive harsh winters, but cattle operations need to store feed for the horses. Thus, the working ranch usually has several barns providing various services for different breeds of livestock.**

OPPOSITE: **Early ranchers who claimed government homestead land of 160 or 320 acres (64.8 or 129.6ha) were required to show that they were working the land. Small, simple houses filled the minimum requirement for claimants who actually lived in town most of the year.**

Provo, Utah, and Santa Fe, New Mexico, or Medford, Oregon, and Porterville, California, can be dangerous, but the barns from all of these areas seem to be easily identifiable as western barns.

In general, we can see that all of these western barns are of relatively simple design and construction. They are smaller than the barns east of the Rocky Mountains, where the interiors often resemble small cathedrals. Western barns are built of redwood or lodgepole pine, woods that weather well and do not need paint.

Other than those on dairy farms, the barns of the West are smaller, since the milder weather allows livestock to remain outdoors all year. The roofs tend to be flatter because the snow load is not a problem in most areas. Cattlemen drive their stock to lower elevations during the winter months.

The earliest California barn structures were made of adobe. Most of them have disappeared, but drawings and a few remaining photographs give us a good understanding of their design. The Californians who first settled the state were cattle ranchers. When George Washington was shipping plows and seeds from England, the first settlers on the West Coast were trading tallow and hides for luxuries from Spain and the Sandwich Islands. Cattle ran wild and grazed on lands without barns or fences.

THE SECRET
WORLD OF
THE BARN

INSIDE AN OLD BARN is a secret, quiet world that few outsiders have ever explored. The late writer Eric Sloane ventured across this frontier and noted in his biographical *Eighty, an American Souvenir* that being alone in the loft of a barn at night was a very tranquil and spiritual experience.

There is an ecology inside a barn, a closed world of insects and animals and plant life that quietly coexist within the barn walls. In this little hierarchy of life, bugs and rodents come to eat and be eaten, these the life and death struggles of very small barn creatures usually ignored by the larger world.

The hay, the grain, the seed, and the animal fodder stored in the barn all attract various freeloaders. Little pests come into the barn with loads of hay, riding along on their own habitat. A warm, dry barn is an attractive shelter for many small animals; some can live in the barn for years without discovery. Spiders are among the most common barn dwellers, encouraged to catch flies as long as their webs are spun in unobtrusive corners.

Barn swallows and owls are welcomed too, since they work hard to keep insects and rodents under control. In some circles, barn owls are thought to be lucky. In the Southwest, bats swoop throughout the evening, eating insects and pollinating the flowers they visit. They, too, are encouraged by farmers who understand their role in the environment.

Mice and rats are frequent residents, hopefully temporary, and farmers have a number of strategies to eliminate them. Some farmers keep a cat to control rodents. A hard-working cat who knows how to hunt is encouraged with regular feeding. Other farmers feel that a cat in the barn can present problems and try to keep them away. Feral cats can be especially troublesome to farmers and are sometimes classified as a varmint, something to get rid of. Any wild animal can carry contagious diseases into the barn, with the potential to infect the domesticated livestock. Poison baits and traps are dangerous to the local residents as well as the unwelcome visitors, so farmers are forced to be cautious.

Keeping a farmyard usually means that the animals—cows, horses, pigs, and sheep—are kept in nearby pens. The manure that is dropped in the yards and the droppings from inside the barn have to be continually hauled away. Animals are susceptible to all sorts of diseases when exposed to unsanitary conditions, and the barn environment can be especially hazardous.

Barns can be a great place to play, especially in rainy weather when there is no room to run in the house. A partly filled hayloft and a rope swing offer opportunities for all sorts of adventures. In the past, barns sometimes functioned as a sort of additional guest room for overnight callers. If there was no room in the house, visitors would sometimes sleep in the barn. For farms who employed a regular "hired hand," the barn was frequently the place for the extra bedroom.

So the next time you see an old barn, remember what a vital place it once was. The ghosts of many creatures, great and small, reside here.

LEFT: **Barn owls might be considered the lords of the barn, the masters of all they survey.**

TOP INSET: **Farm surveys reveal mixed feelings about the presence of a barn cat. About half of the farmers find them hardworking and indispensable; others feel that cats carry disease and create hazards by spooking other animals.**

ABOVE: **The white-footed mouse and other rodents can eat a farmer right out of his profits. Shooting, traps, and poison bait are the conventional solutions for the rodent problem—a hardworking cat is another.**

One of the few remaining adobe barns is in southern California, though time has weathered it considerably. The drawings that remain from the barn on El Escorpion Rancho in southern California portray a structure with many of the internal features of eastern barns. But drawings of the roofline depict the pitch that seems to typify the western barn. Located on property once owned by the San Fernando Mission, it is sited on the crest of a hill with a slope that provides good drainage.

Thought to date to 1796, the El Escorpion barn consists of a two-story adobe structure about 32 by 57 feet (9.8 by 17.4m), between two single-story frame sheds, one on the east side and one on the west. The barn is situated so the large central barn doors face west, in order to take advantage of the prevailing winds of the area. The large doorway is flanked by the two small sheds, which were probably used as a workshop and for wagon storage. The shed on the east side is divided into two sections; one is a sheep fold and the other has three stalls for horses. The central portion of the barn is constructed of adobe brick, two stories high with a coating of lime plaster that has been troweled smooth. It is a substantial building with a stone foundation and has a loft under the eaves. The interior measures 19 feet (5.8m) from the base to the floor of the loft and another 11 feet (3.4m) to the ridge of the roof.

*BELOW: **Horse breeding was a major industry in America at a time when cars were too expensive to own and operate. Today it is considered an expensive hobby, enjoyed by only a few people. The once-common horse barns and stables are not nearly as abundant as they once were.***

Trevarton Ranch Barn

The ninety-year-old log barn on the Trevarton Ranch (also known as the Big Elk Ranch) near Estes Park, Colorado, won an important preservation award in 2000. The barn and its owners were cited by *Successful Farming* magazine and the Barn Again! program of the National Trust for restoring the barn to full functional use. It is a family farm (called a ranch in this part of the world) where beef cattle are a predominant product.

Located at an elevation of eight thousand feet (2438m), the barn was for many years the heart of a mixed farming operation that raised turkeys and chickens for local resorts, and also grew hay and vegetables. In addition, the family had a dairy operation and could house and milk seventy-five to one hundred cows in the log barn. Today, the family grows hay for its horse operation, boarding up to seventy horses during the cold months.

In the tradition of western barns and log cabins, this structure is unpainted. Two stories tall in the center, the barn features doors at each end. The second level is used for hay storage; the lower level was

LEFT AND BELOW: **A prize winner in the Barn Again! program sponsored by Successful Farming magazine and the National Trust for Historic Preservation, the Trevarton Ranch Barn shows that miracles can happen when a heritage barn is lovingly and painstakingly restored.**

RIGHT: *A California barn like this one in the foothills of the Gold Country is a relatively modest structure. Built of redwood, barns here are usually unpainted and survive for years. This structure could be nearly a century and a half old and still be in use.*

BELOW: *Barns of any kind are a welcome sight on the Great Plains, where wheat farms can span thousands of acres. Built as seasonal shelters, these sturdy structures are not as abandoned as they might appear.*

used as a milking parlor. The central barn is flanked by two sheds that stable the horses and run the full length of the barn on each side.

For the recent rehabilitation effort, the logs were given a heavy coat of linseed oil mixed with lacquer thinner. A shiny, new, red steel roof replaced the old, rusted metal roof. It reportedly cost about $20,000 to rehabilitate the 100- by 60-foot (30.5 by 18.3m) barn, far less than the cost of replacement.

The barn is just one heritage building on this farmstead. Companion structures include a granary, ranch house, bunkhouse, and several cabins. This barn and its ranch property are now part of a unique arrangement with Boulder County Parks and Open Space that will allow it to continue to be preserved as a working ranch.

ABOVE: **Western barns were built of Douglas fir or redwood, and left unpainted to weather to a light, silvery gray. The rough-hewn post fence is another western feature; few western farmers paint their fences, even those built of dressed lumber.**

FOLLOWING PAGES: **Despite its somewhat neglected nature, this barn inspires us to stop and admire its appearance.**

The Fountain Grove Barn

Located in the wine-growing region of Napa County, the Fountain Grove Barn of Santa Rosa has always been of great interest to historians. One of the few round barns to appear in California, the Fountain Grove is not a typical western barn, but its design is consistent with its age and location, and its past is so colorful that it reads like a highly imaginative novel.

The Fountain Grove Barn is now on the National Register of Historic Sites, but it came to the attention of historians long before the National Trust registration system was formed. Back in the Depression-era 1930s, when so many Americans were out of work, the government created jobs by building dams and bridges. Artists, writers, and architects were also put to work in their respective fields. One project concentrated on meticulously documenting America's significant architecture. Known as the Historic American Building Survey (HABS), its work continues today. Under this program, the Fountain Grove Barn was photographed and carefully measured, and drawings and diagrams were made. The barn's remarkable history was documented and published at that time.

Like many other round barns, the Fountain Grove Barn was the product of a utopian community. The barn was built in 1875 for Thomas Lake Harris, leader of a group from New York State who called themselves "The Brotherhood of the New Life." The group had supported themselves in New York by making wine, and they transferred their wine-making skills to Santa Rosa.

*BELOW: **A round barn such as this offers the traveler the opportunity to stop and appreciate both the construction and history of such an incredible structure. Found all across America, a round barn usually indicates that the builder was intensely interested in scientific farm practices.***

Thomas Harris created a doctrine from a combination of several religious teachings, though he was primarily interested in Eastern religions like Shintoism. Harris managed to attract the young and impressionable son of a Japanese nobleman to his new cult. The young nobleman, Kanaye Nagasawa, was one of a group of thirty-three students who had been sent to school in England, some of whom became disenchanted and sought to escape. One such student found asylum in America with the Brotherhood, and when the group moved to California in 1875, Nagasawa, then twenty-two years old, went West.

Two wine makers helped establish the Fountain Grove winery complex, a Missouri viticulturist named Dr. John Hyde and young Kanaye Nagasawa. The Harris group initially acquired 2000 acres (810ha) of land and set up the winery as their economic base. Gaining an international reputation, the winery was shipping wine under its Fountain Grove label to outlets in London and New York by the 1880s. In 1892, however, the little colony was rocked by sex scandals, and Harris and his wife fled California. The disciples were then left to fend for themselves, and the winery operation remained under the direction of Hyde and Nagasawa. Gradually, the wine operation was taken over and managed by Kenaye Nagasawa alone. In 1900 the property was sold to the remaining five members of the Brotherhood by Harris's heirs, and Nagasawa eventually inherited the entire property.

Known as the "the Japanese baron of Fountain Grove," Nagasawa reportedly smoked cigars, wore tweed suits, and spoke English with a Scots burr. He never married, but brought his nephews to America and lived to the age of eighty-one.

The Fountain Grove Barn was built to house the horses used in the winery operation and for general rural transportation. The barn is about 70 feet (21.3m) in diameter and just under 60 feet (18m) high when measured to the top of the cupola. Drawings show a structure fairly typical of round barns, with a central hay loft on the upper story and stalls for the animals on the first floor. The barn was built into the side of a hill, enabling the manure to be gathered and hauled away on the lower level.

ABOVE: *While most round barns appear in the Midwest, a few are scattered across the West as well. Nearly all of these barns are more than a century old, most built during the time of frontier expansion.*

The redwood barn is noted for its interesting open cupola, which features open latticework that resembles a large basket and whose purpose is to provide ventilation. The stable floor has decorative paired windows; the upper floor has an exterior opening to the hayloft but no other openings. The conical roof was originally made of redwood shingles but was covered over by tar paper in later years.

The Koster Barns

The Koster ranch represents a typical California farming operation of the 1880s, when wheat ranches covering thousands of acres provided the economic base that helped open the vast San Joaquin Valley for development. The Koster farm and its two barns are located in the western San Joaquin Valley. The farm was homesteaded in the 1880s by the Koster family, and today the fourth, fifth, and sixth generations of the Kosters live on the farm.

The family's farmhouse burned down many years ago and was replaced by a single-story home, but the barns, both still in use, are original, and are over one hundred years old. They hold all of the old equipment: The hitches and the harnesses once used in the farm operation, the old wagons and surreys, the plows and planters. Simply walking into a Koster barn becomes a lesson about the agricultural practices of California a century ago.

*OPPOSITE: **A barn represents a considerable initial investment, so keeping a barn in good repair is an effort that pays off over a long period of time. With a good roof and a little paint, a simple barn such as this can continue to contribute significant returns for years to come.***

*ABOVE: **The Koster farm near Tracy, California, has two nearly identical livestock barns, one for horses and one for mules. The barn is a century old but the Holt Crawler is a little newer—it's World War I vintage.***

Just in case someone on horseback visits, the hitching post is still in place in front of each barn. The watering trough is there too. All of the traditional comforts are still available from the days when horse

and mule power supplied the muscle to operate the threshing machines, though the horses and mules were sold off in 1914 and replaced by a Best 25 tractor that pulls the Harris harvester. But the harnesses were kept in the barns, just in case they would ever be needed again. Having a large barn with plenty of storage space means that nothing useful ever has to be thrown away.

The Koster farm was originally a grain operation for growing wheat, oats, and barley, but modern irrigation has made tree crops such as almonds, walnuts, and apricots much more profitable to grow. The barns are still used to house farm equipment and to shelter the family's pets—animals that sometimes include a horse or two or perhaps a small herd of sheep.

Duvenek Barn at Hidden Villa

Josephine and Frank Duvenek were one of the earliest families in California to preserve their farm by turning it into a nonprofit educational center. The farm is located on 1600 acres (648ha) of the coastal foothills of Los Altos, near San Francisco, and the educational program brings about twenty thousand school children each year to spend a day on the farm and learn about the "interconnectedness of all living things." Much of the Duvenek farm has been a wilderness preserve for about fifty years, but the nonprofit educational center was established in 1970.

The barn is a modest California barn, built of weathered redwood with a coat of white paint on its aged surface. The central portion is two stories; like typical western barns, it has a pair of long sheds running the entire length of the barn. The farm runs year-round as a completely organic operation. For many urban students, this is their first opportunity to gather eggs, see a cow milked, and plant a garden.

Although many communities have urban farms with displays for local children, the Hidden Villa experience offers students the chance to get their hands dirty. Students visit the farm over time to see how the environment changes through the seasons.

PREVIOUS PAGES: **Barns somehow represent our strengths, beliefs, and national attitudes. As our barns disappear from the landscape, will our heritage evaporate too?**

ABOVE: **It should be easy to tell what type of farm operation this represents just by analyzing the buildings. It's probably a cattle ranch rather than a farm— there is no silo but there is a large hay barn and a horse barn.**

OPPOSITE: **Some barns were built entirely by local farmers, and some were built with kits ordered from Sears, Roebuck, or other agricultural specialists. Both methods produced beautiful and durable structures.**

Old Barns on a New Frontier

Farming is a dwindling occupation in this techno-logical society, and urban sprawl is rapidly gob-bling up thousands of acres of farmland across the nation. Because of this, barns are disappearing from the American landscape. Left alone, though, barns will survive for decades. The mild climate in the West and Southwest is gentle on buildings, especially when the structure is built of old-growth redwood and is on a good foundation.

OPPOSITE: **Wind and weather have taken their toll, and now this unique barn looks like a good candidate for a restoration project. The combination of a cross-gabled roof with the hipped sheds on each end are uncommon, and the ventilation dormer suggests that the barn could have been built around 1890.**

ABOVE: **The double doors frame a spectacular western view, but the door placement suggests that this is an early threshing barn. If so, it deserves very special consideration as a rare structure, since most threshing barns were built east of the Rocky Mountains.**

LEFT: **Benign neglect has probably preserved many barns that would otherwise have suffered from the so-called improvements of somewhat misguided handymen. Modifying this structure to suit modern convenience would have destroyed its integrity.**

BARN AGAIN!

SEEING A DETERIORATING FARMSTEAD is a very sad experience. Even if you were not raised on a farm, an empty and abandoned house and barn are somehow very disquieting. So it is especially heartwarming to realize that many institutions are now making a concerted effort to maintain our agricultural heritage. At this time there are several programs that address the special problems inherent in barn preservation.

One of the most successful barn preservation programs was originated by the National Trust for Historic Preservation, a national organization that protects America's old buildings. Chartered by Congress in 1949, the trust is a private, nonprofit organization that fights to save buildings especially significant to our American heritage. Sometimes thought to be obsolete for modern farming requirements, barns were especially vulnerable to demolition. The trust understands that preservation is only part of the solution: Barns need to be used and valued in order to survive.

The Barn Again! program was initiated in 1987 to provide advice and information to barn owners about cost-effective techniques for saving their barns. Technical guides, practical information, financing options, and workshops are all part of a strategy to help farmers get the most out of their old barns.

Many old barns are functionally obsolete. The trust offers a number of economical strategies to allow an old barn to accommodate the larger equipment that farmers use today. After nearly fifteen years, the trust can point to scores of examples where barns have been rehabilitated for use. These case studies are a continuing inspiration for others who face the same dilemma.

The Barn Again! program is jointly sponsored with *Successful Farming* magazine, the largest paid-subscription farm journal serving farm families. The magazine sponsors an annual contest to honor barn owners who have successfully rehabilitated their barns.

The Barn Again! program is headquartered at the regional office of the National Trust in Denver, Colorado. Contact the organization for more information:

Barn Again!
National Trust for Historic Preservation
910 16th Street, Suite 1100
Denver, CO 80202
(303) 623-1504
www.barnagain.org

In addition to the national Barn Again! program, many states have also recognized the importance of saving the barns that contribute to their unique heritage. These state organizations include a wide number of groups across the country, such as the Wisconsin Barn Preservation Initiative. Contact your local preservation organization to find a barn preservation group in your area.

Dozens of barns have been restored and put to productive use as...barns. While thirty years ago it was fashionable to make barns into houses, today we are finding them honest work once again.

The best way to preserve the vintage barns that remain is to appreciate them, to document their history, and to share information. It is also important to find a good purpose for the barn in question, since having the building occupied will keep it alive. Turning a barn into a house is sometimes the only viable option, but it usually destroys the integrity of the structure and limits its future possibilities.

Appreciating vernacular buildings such as our old barns helps us to understand the processes that shape the communities that surround us. It is said that a person with no past has no future. So our vintage barns are an important asset, helping us provide for our future by reminding us of our past.

LEFT BOTTOM:
Vernacular architecture, unpretentious structures built of local materials, helps provide a strong regional identity to areas whose unique character is threatened by monotonous standardization.

Glossary

A

Amish: a conservative Mennonite religious group, followers of the teachings of Jakob Ammann, a seventeenth-century Swiss bishop

B

barley: a widely distributed cereal plant, useful as a grain but also an important ingredient in making beer, ale, and whiskey

barn: an old Middle English word originally describing a structure used to store barley

bent: a transverse frame designed to carry either vertical or horizontal loads

byre: pronounced bire or sometimes bye, this is a shed, usually open on one side and frequently adjacent to the barn, that shelters livestock in the farmyard

C

chaff: the light, brittle husks of a kernel of grain that become separated from the kernel during the threshing process

corn: Americans usually use this term to describe maize, a New World grain, but European farmers have used the term for centuries to refer to their local grain

crib: a small building with tall sides used to store corn on the cob; the crib walls are usually slatted to encourage the air circulation that dries the corn

cupola: a small structure appearing on top of a roof, serving as a belfry or a ventilator

D

Dutch barn: built by the Dutch in America in the 1600s, this type of barn used a style of architecture unknown in the Netherlands, with a steeply pitched roof. (Artist-historian Eric Sloane notes that a Dutch barn is actually the most European of American barns.) There is some understandable confusion in terminology since the word "Dutch" is frequently used in place of "Pennsylvania Dutch," meaning German (Deutsch). A Pennsylvania Dutch barn refers to a style of barn architecture favored by Amish and Mennonite farmers in Pennsylvania

F

flail: a tool, consisting of a long stick attached to a shorter stick that swings freely, used to thresh grain by hand by knocking heads of grain from their stalks

fork: one of the most universal barn tools, the hay fork, or the pitchfork as it is sometimes called, is used to throw the hay into the loft. Forks are made with three or four tines. The three-tined fork is to pitch hay; the four-tined fork is to pitch manure—don't mix them up!

G

gambrel: a roof shape originally peculiar to barns and commonly referred to as a barn roof. In reality, the term gambrel refers to the hock of an animal and recalls that peculiar bent shape of the lower leg, as exists in horses

granary: a storehouse for grain after it has been threshed and husked

H

hay: any grass, clover, or alfalfa of suitable nutritional value, used as a feed for livestock. Hay is sometimes confused with straw by unknowing city folk.

hay carrier: a device consisting of a large fork and pulley mounted at the end of the ridge of a barn roof, which enables the farmer to unload hay from the wagons and store it in the upper reaches of the barn

hay fork: sometimes called a pitchfork, this is a common farm tool used to throw, or pitch, hay into a wagon or into the haymow

hay mow: the place in the barn where hay is stored

hex sign: a stylized hexagonal (six-sided) decorative symbol, once reportedly used for protection against evil spirits. Many late nineteenth-century barns feature decorative hex signs

L

loft: a space just under the roof, used for storing hay

M

manure: any substance used to fertilize and enrich a crop

Mennonite: a Christian body of religionists who take their name from sixteenth-century bishop Menno Simons. Persecuted for their faith, the first colony of Mennonites settled in Germantown, Pennsylvania, in 1633. They are noted for their plain dress and simple living, and conservative sects follow traditional farming practices

milo: a grain sorghum introduced into the United State in the nineteenth century and widely used as a feed for livestock

P

Pennsylvania Dutch: the descendants of a group of eighteenth-century immigrants from southwest Germany who settled in Pennsylvania. "Dutch" is a corruption of "Deutsch," which means German

piggery: a shed and area where pigs are kept. Generally, it is different from a sty, which refers to the open yard or pen used only for swine

R

rye: a widely cultivated cereal, also used in making some types of whiskey

S

seed bin: a storage area for the best of the crop, used in sowing the next season's crop

Shakers: a religious group from England who first settled in America in 1774. Advocating celibacy and common ownership of properties, they are noted for their exquisite design of agricultural tools and farm implements. They are also noted as proponents of the round barn

sheaf: a bundle or cluster of cereal grains. After cutting, the crop is gathered into sheaves for threshing

silo: a tall and distinctive farmyard structure used to store silage, or feed, for the livestock

stanchion: an upright bar or post at the end of a stall, used to confine a cow's head during milking

straw: the discarded stalk of grain after it has been threshed. Straw is a very useful commodity and is used for animal bedding and for mulch

sty: a pen or enclosure for swine, usually kept some distance from other farm structures because of the odor

T

threshing: pronounced "thrashing," refers to the process of separating the kernels of grain from the stalks. In early America this was done by hand. Inventors such as Jerome Increase Case developed mechanical threshers as early as the 1850s, making the American farmer extremely productive

threshing floor: the upper floor of the barn, originally a large covered room used to thresh, or beat, the grain to separate the heads from the stalks

W

wattle and daub: mud and straw, or plaster and reeds, or any similar combination of the two components that were used as construction materials to cover the walls in early farm structures

weather vane: device mounted on the ridge of a barn roof that moves freely with the wind and indicates which direction the wind is blowing. A smart farmer would situate his barn and weather vane so that they would be visible from his kitchen window

Y

yoke: a device for joining a pair of draft animals, usually oxen. The word is frequently used to denote a pair of working oxen

Selected Bibliography

Arthur, Eric, and Dudley Witney. *The Barn: A Vanishing Landmark in North America*. Boston: New York Graphic Society, 1972.

Halberstadt, April. *Farm Memories: An Illustrated History of Rural Life*. Osceola, Wis.: Motorbooks International, 1996.

Halberstadt, Hans. *The American Family Farm*. Osceola, Wis.: Motorbooks International, 1996.

Hartley, Dorothy. *Lost Country Life*. New York: Pantheon Books, 1979.

Leffingwell, Randy. *The Great American Barn*. Stillwater, Minn.: Motorbooks, 1997.

Rasmussen, Wayne D., ed. *Readings in the History of American Agriculture*. Urbana: University of Illinois Press, 1960.

Sloane, Eric. *An Age of Barns*. New York: Funk & Wagnalls, 1972.

Web Sites

Barn Again!:
www.agriculture.com/ba/aboutba!home.html

The Barn Journal:
http://museum.cl.msu.edu/barn/

Barn Owls:
www.barnowl.org/uk

The Dutch Barn Preservation Society:
www.schist.org/dbps.htm

National Trust for Historic Preservation:
www.nthp.org/

Round Barns:
www.vreug.com/barns

Wisconsin Barn Preservation Program:
www.uwex.edu/lgc/barns/barns.htm

Index